ONE SMALL SQUARE®

Backyard

by Donald M. Silver

illustrated by Patricia J. Wynne

LEARNING
TRIANGLE
PRESS

Connecting kids, parents, and teachers
through learning

An imprint of McGraw-Hill

New York San Francisco Washington, D.C. Auckland Bogotá
Caracas Lisbon London Madrid Mexico City Milan
Montreal New Delhi San Juan Singapore
Sydney Tokyo Toronto

The name and picture of every plant and animal in this book can be found on pages 38-43. If you come to a word you don't know or can't pronounce, look for it on pages 44-47.

Dedication

For Ole Risom

—for making sure there were as many books about earthworms as there were about Lowly Worm.

Thanks to

David and Jason Silver for helping explore the backyard in winter and summer; Thomas L. Cathey and Maceo Mitchell for caring for the animals; and Ivy Sky Rutzky and Karen Malkus for their enthusiastic suggestions, comments, and insights.

Library of Congress Cataloging Number 97-5706
ISBN 978-0-07-057930-9
MHID 0-07-057930-X
19 LWI 22

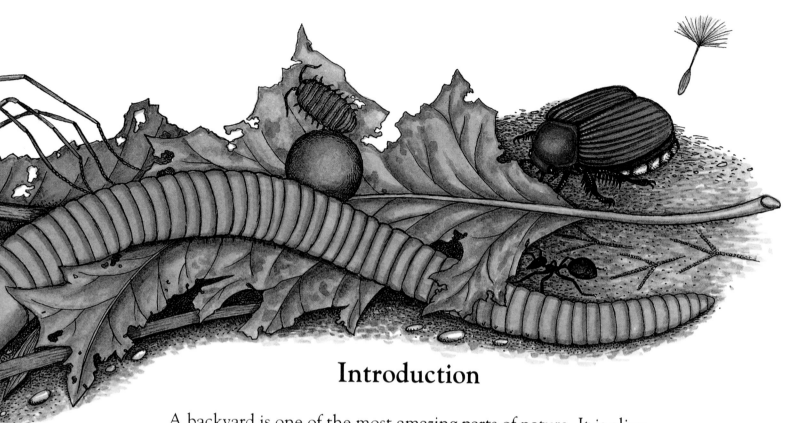

Introduction

A backyard is one of the most amazing parts of nature. It is alive with creepers and crawlers, lifters and leapers, movers and mixers, munchers and scrapers, singers, buzzers, chirpers, climbers, builders, buriers, and recyclers.

In a backyard, hungry hunters search for clever hiders that will do just about anything not to become the next meal. Food makers capture energy that speeds to Earth from the sun across 93 million miles of space in only 500 seconds. Colorful petals double as landing strips for delicate butterflies and hardworking bees. And that bug on stilts, daddy longlegs, zips over leaves and twigs with the skill and speed of an acrobat.

Go outside and catch a dandelion seed zigzagging to the ground. Pick up a rock that crumbles in your fingers. Look for tracks left in mud by an ovenbird. Listen to a cricket rub its wings together. Watch a robin tug on an earthworm holding on to earth for its life.

Explore just one small square of a backyard—your own or someone else's—and you will uncover clue after clue about how nature works. Put all the clues together and you will be able to figure out how living things all over the Earth are connected.

This book is about discovering what's happening in your backyard. As you explore your square, the tools shown on this page will come in handy. With a magnifying glass, you will enter the world of the tiny. With a small shovel or a trowel, you can dig for clues hidden underground. With a collecting jar, you can look closely at an earthworm before returning it to the soil. Always carry a notebook and a pencil or a pen so you can write down what you find and draw pictures of animals and plants you want to learn more about later. When your notebook is complete, friends and family can use it as a guide for their own exploration of your backyard.

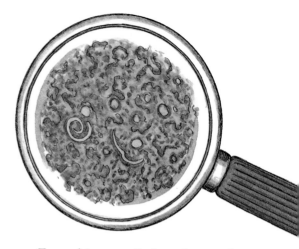

Everything you find can be seen larger with an inexpensive magnifying glass. You will discover tiny creatures you didn't know were there and exciting details that will amaze you.

Use gloves to protect your hands, a pan to hold things, and a brush to gently sweep away dirt. Spread out newspaper to hold the soil you dig up, and always put the soil back later. You may come up with other tools. Note in your book what they are and how they helped.

One Small Square

Listen for the clear song of the hermit thrush.

The grass in your backyard is home to many small animals. Look for grasshoppers with long jumping legs, woolly bear caterpillars, and garden spiders hanging from silk they spin.

You share your backyard with all the plants and animals living there. They depend on your yard for everything: food and water; space to grow; and places to hide, sleep, and build nests. That's why it is very important not to disturb their homes, harm them in any way, or move them to a spot where they can't protect themselves. So, instead of looking into every hole and corner of your yard, choose the part you'd most like to explore. Take along a yardstick or meterstick (about three inches longer) and use a twig to draw a square in the earth the length of the stick on each side. Show an adult your square and ask if it is okay for you to work and dig there. Promise to dig down only about 18 inches and to leave the yard in good shape.

Follow along as a square like the one in the picture is explored. There will be activities you can do in your yard. Remember: not all of the animals and plants in this book will be in your yard. But no matter where you live, you will turn up similar kinds of creatures, just as wonderful.

Plant-eating aphids make a tasty meal for a hungry ladybug beetle.

Billions of living things in the soil are so tiny that you need a microscope to see them.

When pill millipedes sense danger, they curl into balls. Can you find some in the picture?

A springtail can't fly. When it unhooks the "tail" folded under its body, this insect springs into the air for a quick escape. You might be lucky enough to catch one in your collecting jar.

As a roundworm moves underground between soil particles, it is trapped by fungus threads.

7

Nearly all the plants in your yard grew from seeds made by flowers. To make seeds, pollen from male flower parts (stamens) must reach female parts in the pistil.

Stamen

Pistil

Watch closely as a hummingbird hovers, or a butterfly or a bee lands on a flower. Like you, they are attracted to the bright colors, beautiful patterns, and wonderful odors of flowers. But also they are hungry for the sweet-tasting liquid nectar that flowers make.

As the animals drink, they brush against stamens. Pollen grains cling to their bodies. When they fly to the same kind of flower, some of the pollen grains drop and land on female flower parts. This is called pollination.

Floating on a breeze, feathery-looking dandelion seeds may drift into or out of your square.

Pollen

A butterfly sips sweet nectar through its long mouth tube.

As a honeybee collects nectar for making honey, powdery pollen sticks to its body. The bee packs the pollen into lumps on each back leg. Then it flies to another flower or returns to its nest.

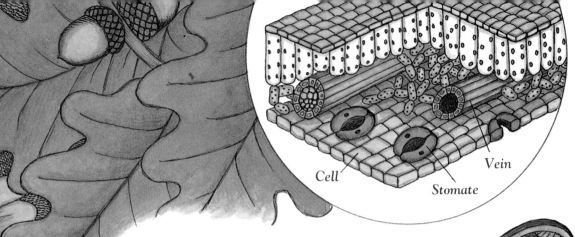

Cell

Vein

Stomate

A leaf has just a few cell layers. Food is made in cells just below the leaf surface. Parts of these cells contain green chlorophyll that traps light energy. Leaf veins deliver water from the roots. Gases move in and out of leaves through tiny openings called stomates.

The way plants make food is called photosynthesis. Photo means "light," and synthesis means "putting together with."

The Food Makers

Without smokestacks or machines, without assembly lines or wires, without even a sound, plants capture energy from the sun. They use that energy to make sugar from water in the ground and carbon dioxide gas in the air. When they are done, they have oxygen gas left over. They release it into the air.

All the plant leaves in your small square are making sugar. They use that sugar to make other nutrients—starches, fats, and proteins—so they can grow and stay healthy. They are also giving off oxygen that you and the creatures in your square breathe to stay alive.

What's Living in Your Backyard?

Plants

Animals

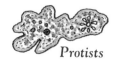
Funguses

Protists

Monera

9

Your Backyard Notebook

Whenever you work in your square, record in your notebook the date, time, and weather. Note what each animal is doing, what it is eating, or where it is hiding. Draw spiders spinning webs and caterpillars nibbling leaves.

Red bug eating smaller green bugs. 4:00 pm

Use More Than Your Eyes

Sit in your square and close your eyes. What sounds can you hear? What do you smell? Sounds and smells are important clues to which birds are visiting, if the wind is blowing, or if the soil is damp. Write them down.

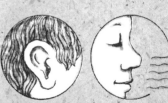

Sparrow hawk

A bird's beak helps it get the kinds of food it wants to eat. A sparrow hawk's strong, hooked beak tears into animal flesh. The beak of a white-throated sparrow cracks seeds. The sharp, pointed beak of the black-throated blue warbler is perfect for capturing insects.

White-throated sparrow

Black-throated blue warbler

Blackpoll warbler

If necessary, a gray squirrel will hang upside down from a low branch to reach an acorn in your square. Baby squirrels drink their mother's milk until they are old enough for seeds and nuts.

Baby squirrels

The brown thrasher spends most of its time searching for insects on the ground. But it may fly into bushes.

Your yard may serve as a food or rest stop for birds that are migrating—traveling between their winter and summer homes. They often stay just a day or two and return only when the seasons change, if at all.

10

Guests for Dinner

You don't have to put up an EAT HERE sign in your square to invite squirrels, chipmunks, and robins to visit. Animals will drop by as long as they see, smell, or hear the kind of food they love to eat.

Animals fear predators—other animals that would like to eat *them*. They may think you are one. So sit away from your square and be as quiet and as still as you can.

Before long, a squirrel may snatch an acorn or a sparrow may grab a seed. These plant eaters get the nutrients they need for energy and growth from plants.

Meat eaters couldn't care less about what's growing in your square. When they see the kind of animal they eat, they get ready to pounce on it. When one animal eats another, nutrients are passed along.

If you live where there are birds of prey such as hawks or eagles, you may see a bird suddenly swoop down and carry off the sparrow that was eating the seed. This is nature's way. Plants make food. Animals eat and are eaten.

Flight feather

Contour feather

Downy feather

Most birds build nests hidden in trees, safe from predators. If you find a nest in a bush that has eggs or baby birds inside, do not touch it. You don't want to crack an egg or lead a hungry predator to the nest.

Your Backyard Notebook

Start a guest list of plant eaters and meat eaters. Keep adding to it as you continue to explore your square. How does the list change with the seasons?

Who's There?

There are so many kinds of living things that you won't be able to identify every one you see. Adults need help, too. Try a field guide—a book with the names and pictures of animals or plants you are most likely to find where you live. To make identification easier, a guide points out features such as shape, size, color, patterns, and sounds. There are field guides for birds, insects, wildflowers, and leaves— for just about everything. Use the guides you have at home or some from the library.

When you draw a picture in your notebook, mark down as many features as you can. The more you know about what you see, the more likely you will be able to find out what it is.

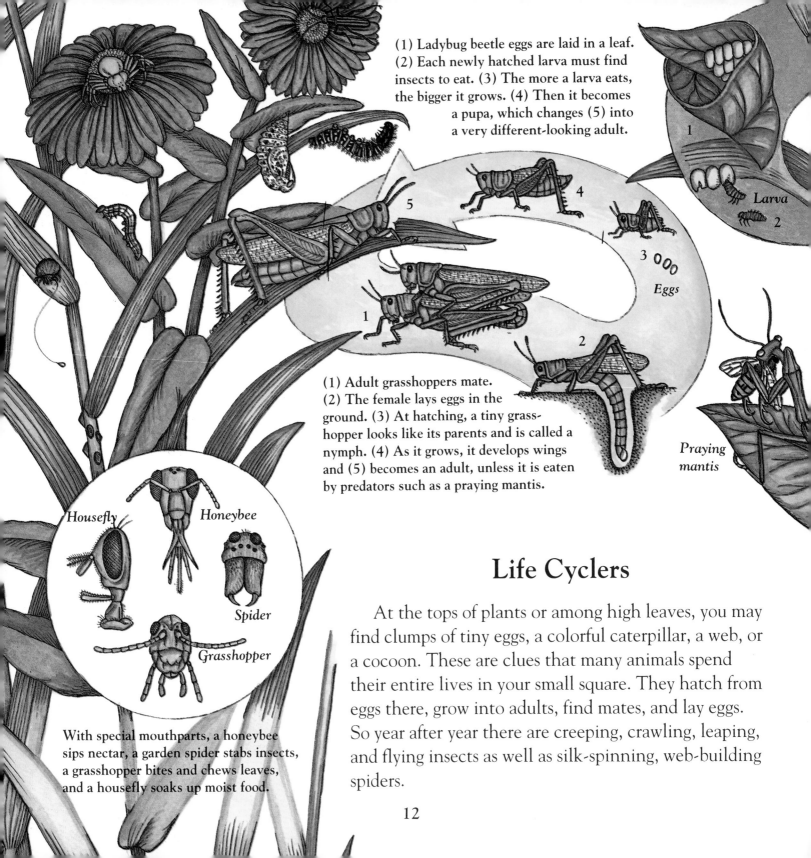

(1) Ladybug beetle eggs are laid in a leaf. (2) Each newly hatched larva must find insects to eat. (3) The more a larva eats, the bigger it grows. (4) Then it becomes a pupa, which changes (5) into a very different-looking adult.

Larva

3 000 Eggs

(1) Adult grasshoppers mate. (2) The female lays eggs in the ground. (3) At hatching, a tiny grasshopper looks like its parents and is called a nymph. (4) As it grows, it develops wings and (5) becomes an adult, unless it is eaten by predators such as a praying mantis.

Praying mantis

Housefly

Honeybee

Spider

Grasshopper

With special mouthparts, a honeybee sips nectar, a garden spider stabs insects, a grasshopper bites and chews leaves, and a housefly soaks up moist food.

Life Cyclers

At the tops of plants or among high leaves, you may find clumps of tiny eggs, a colorful caterpillar, a web, or a cocoon. These are clues that many animals spend their entire lives in your small square. They hatch from eggs there, grow into adults, find mates, and lay eggs. So year after year there are creeping, crawling, leaping, and flying insects as well as silk-spinning, web-building spiders.

5

Aphid

Pupa

4

3

Garden spider

Egg case

(1) The common blue butterfly lays eggs on (2) a clover leaf. (3) Each hatched caterpillar gives off sweet liquid honeydew that little black ants drink. In return, the ants bite any predator that gets close. (4) The fully grown caterpillar changes into a butterfly inside a hard case called a chrysalis.

All the different stages of a living thing make up its life cycle. The human life cycle includes being a baby, a child, a teenager, and an adult, and going through old age. A moth hatches from an egg as a many-legged caterpillar that eats and eats and grows and grows. At this stage it is called a larva, a term used for a very young animal that doesn't look at all like its parents. When fully grown, the caterpillar stops eating and spins a cocoon around itself. It is now called a pupa. The pupa changes into an adult that leaves its cocoon as a moth with wings and six legs. It is ready to mate.

13

Insects and Other Bugs

None of the animals on these pages will bite or sting you. Try to look at them with your magnifying glass.

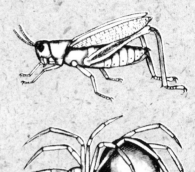

If you find a spider and a grasshopper, compare the two. Which has antennas and wings? The number of legs a grasshopper has is the same number all adult insects have. The number a spider has is the same as a daddy longlegs, a scorpion, a mite, and a tick.

What Will It Eat?

Dissolve sugar in water and place it in a plate or saucer in your square. Nearby put a small piece of raw meat, such as hamburger, and a piece of lettuce. Move away and list in your notebook which animals go for the sugar, the lettuce, and the meat. Be careful! Your picnic will attract stingers and biters.

meat

sugar water

lettuce

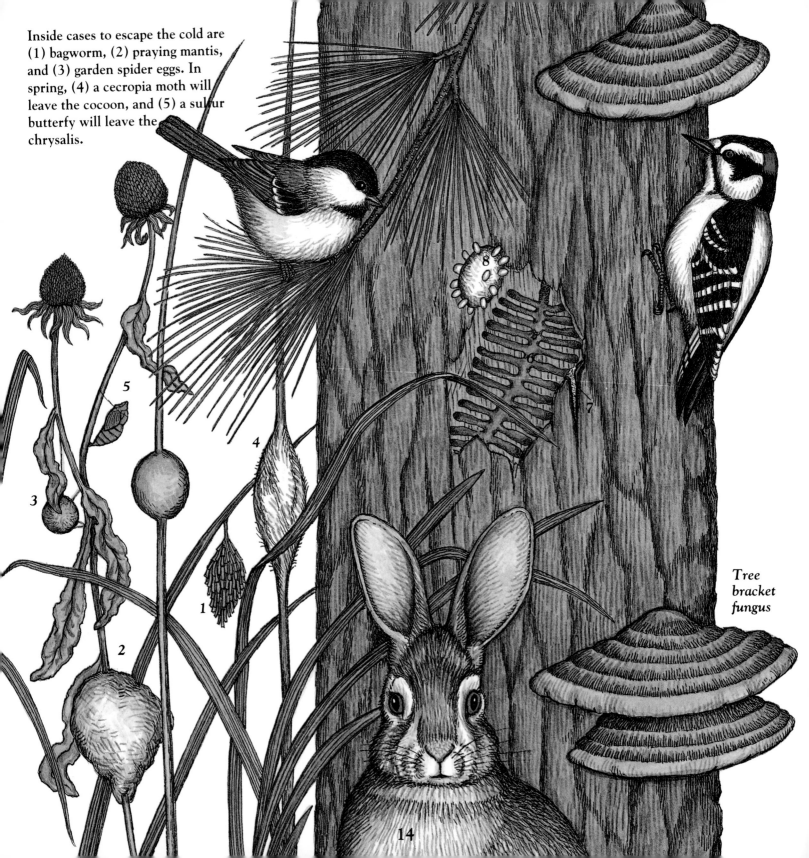

Inside cases to escape the cold are (1) bagworm, (2) praying mantis, and (3) garden spider eggs. In spring, (4) a cecropia moth will leave the cocoon, and (5) a sulfur butterfy will leave the chrysalis.

Tree
bracket
fungus

14

Out in the Cold

On the first cold day of the year, out come your scarves and gloves, sweaters and hats. But for backyard plants and animals, it may be too late to protect themselves from the harmful effects of frost. To avoid being weakened or killed, they have to prepare ahead for the time when water turns to ice.

Visit your small square on a winter day. Look for signs of how each living thing survives the coldest months of the year. Most plants shed their green leaves and live off food they stored in stems and roots during the warmer months. That plump rabbit searching for moss put on extra body fat and grew thicker fur months ago to keep out the chill. Birds that didn't fly off to warmer places fluff their feathers to prevent body heat from escaping. If they can't find water, they may have to drink drops dripping from icicles.

Few, if any, insects can be found in your yard. Most are dead, but they have left eggs on plant stems and tree trunks. Protected by cases or hidden under bark, the eggs hatch when spring returns.

Look for (6) tunnels bored in bark by a female engraver beetle for her eggs, (7) a grass spider's egg case beneath loose bark, and (8) braconid wasp eggs laid on a tent caterpillar paralyzed by the wasp's sting. When the wasp eggs hatch, the caterpillar becomes food for the young insects.

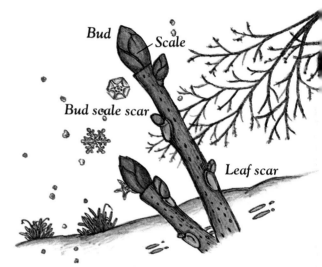

If you see a patch of something black like pepper moving on top of the snow, it is really a herd of thousands of snow fleas. Using your magnifying glass, can you tell that snow fleas are springtails?

Bud *Scale*

Bud scale scar

Leaf scar

In a bud are plant cells that will grow in spring. They are protected from wind and cold by scales. Look for scars left by fallen bud scales and leaves. Then measure the distance between two bud scale scars. That's how much the plant grew in a year.

Gall

Larva

A gallfly lays eggs on the stem of a black-eyed Susan. When an egg hatches, the larva tunnels into the stem. It gives off a chemical that makes stem cells grow into a ball called a gall. Inside the gall the larva is safe for the winter. It feeds on the gall, eating its way out. Watch for galls on stems and trees in your yard.

Fence lizard

A fence lizard lies in the sun, soaking up heat to get its muscles working. A five-lined skink looks for insects.

Five-lined skink

A chipmunk has packed its cheek pouches with seeds to store in its burrow. It pauses to make sure no predators are near.

If the robin loses at tug-of-war with the earthworm, it stays hungry. If the worm loses at tug-of-war, it is eaten.

To you this is a rock. But to this vole it is a dining room complete with seeds to eat.

By flicking out its tongue, a garter snake can pick up odors of frogs, toads, and mice it will swallow whole. Never kill a garter snake. It will not hurt you.

The Eastern box tortoise hasn't even swallowed the slug in its mouth, yet it is looking for more to eat.

Bombardier beetle

The bombardier beetle saves itself by spraying hot chemicals at the sticky tongue of an American toad.

Making a Living

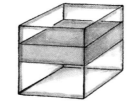

By the time you are on your way to school and adults are on their way to work, most of the animals in your small square are already busy "making a living." Without supermarkets and refrigerators, they have to find enough food every day to provide the energy they need to run, jump, climb, or fly.

There are no weekends off, holidays, or vacations for backyard animals. For them, danger is always present. While an animal warms up in the sun or cools off in the shade, a predator may suddenly appear. There will be only a split second in which to escape. Whether building a nest, digging a burrow, munching leaves, or caring for its young, an animal had better be on guard for movements, sounds, and odors—or else!

Making their living within about six inches of the ground are lizards, chipmunks, toads, and other animals that have skeletons made of bone. In the same neighborhood there are many, many kinds of boneless animals, including snails, spiders, beetles, ants, and other insects.

Any bone you find in your square belonged to an amphibian (such as a frog), a reptile (such as a turtle or snake), a bird, or a mammal. All bony animals are vertebrates. Their backbones are made up of many small bones called vertebrae.

Except for the earthworm, beetle, and slug, all the animals on these pages are vertebrates.

Skull

Every morning and evening, visit the square and note in your book anything left by animals. A nut split down the middle may be a sign that a squirrel was there. Scratch marks were probably made by a cat. Start a list like this one:

which bird lost a feather?

nut opened by a squirrel?

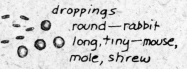

droppings
round—rabbit
long, tiny—mouse, mole, shrew

did the cat that scratched find the mouse that chewed?

a mole lives here

Fresh droppings mean that an animal was just there. Remember that droppings decay over time, and the nutrients in them are returned to soil to be reused.

If the jumping spider misses the moth, its silk safety line will keep it from hitting the ground.

Safety line

Each kind of animal lives where it does because it is adapted, or fitted, to survive there. Be on the lookout for a chipmunk stuffing its mouth pouches with seeds. Watch a spider building its web between leaves. Try to spot a snail pulling into its shell as a robin lands nearby and a tiger beetle seizing a caterpillar with its powerful jaws. See if you can find a fence lizard with colors and patterns so well blended into a rock that it doesn't seem to be there at all. These are all examples of how body parts and movements work to help animals stay alive.

Most animals lay lots of eggs each year. If all the eggs hatch, there will be many, many young competing for the same food and the same space. Because there is never enough of either for every creature, not all survive. Those best adapted to making a living in your square have the best chance of growing, escaping predators, finding a mate, and laying eggs.

Scale insect

Tiger beetle

On a single clover plant, a sulfur butterfly sips nectar, a sulfur caterpillar chomps at leaves, and scale insects suck juice from the stem. Does the caterpillar sense that a tiger beetle is about to attack?

Little black ants leave a chemical trail for other ants to follow from their nest to the newfound food.

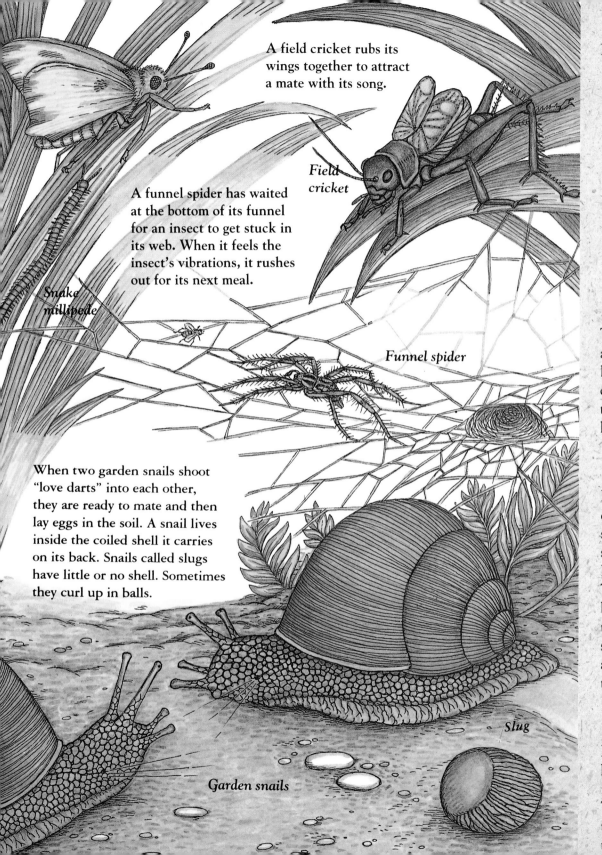

A field cricket rubs its wings together to attract a mate with its song.

A funnel spider has waited at the bottom of its funnel for an insect to get stuck in its web. When it feels the insect's vibrations, it rushes out for its next meal.

Field cricket

Snake millipede

Funnel spider

When two garden snails shoot "love darts" into each other, they are ready to mate and then lay eggs in the soil. A snail lives inside the coiled shell it carries on its back. Snails called slugs have little or no shell. Sometimes they curl up in balls.

Slug

Garden snails

Who Did It?

Look for leaves like these:

sap sucked from leaf

ragged leaves

fine web

tiny holes in leaves

They are clues to the kinds of animals in your square. Draw the leaves you find. How do they compare with these? Do you think the same kind of animal left its mark in your leaves?

Web Trick

Each kind of spider builds its own kind of web. Some webs are shaped like wheels. Others are shaped like triangles and funnels. Draw each web in your square. Then try to find the spider that built it. If you touch a web gently with a blade of grass, the spider will feel the vibrations and may come out into the open to check for a trapped insect.

Hurt Animals

If you find an animal that is hurt, do not go near it or touch it. Call an adult to take a look. The adult can contact people trained to handle, care for, and treat injured creatures.

Definitely Not a Golf Ball

One summer morning, after a night of rain, you may discover what looks like a golf ball in your small square. Leave it alone. By the next day it may be the size of a baseball. Later it may be larger than a basketball. Just when you think it is going to be as big as a watermelon, it starts to shrink. A hole opens in the top. Tap the ball. Out puff thousands and thousands of tiny brown specks. The strange puffing ball is a mushroom—a puff-ball mushroom. The specks are its spores, cells from which new mushrooms grow.

Most of the mushrooms in your yard are just a few inches tall and look like umbrella-shaped plants. But they don't have chlorophyll to capture light energy from the sun. They can't make their own food, grow

For most of the year a mushroom lives as thin threads underground. When threads from two mushrooms join, a ball forms and grows a stalk with an umbrella-shaped cap. Some new mushrooms may grow from the millions of spores shed from under the cap.

Plant stems do much more than support and hold leaves up to light. Inside a stem are tubes called xylem and phloem. They carry water and minerals from the roots to the leaves and carry food down from leaves to the rest of the plant.

Xylem

Phloem

Horsetail

Moss

Plants compete for light, water, soil, and minerals. Not all have roots, stems, and leaves. Moss has no roots. Horsetails lack leaves. Neither grows flowers.

flowers, or produce seeds. They belong to the group of living things called funguses, which also include molds and yeasts. Many funguses get food by absorbing it from the remains of dead plants and animals. When they break dead matter down so it can be reused, they are acting as nature's recyclers. (Remember: never eat growing mushrooms, for some are poisonous.)

The damp places where mushrooms grow are great to look at through your magnifying glass. Take a close look at the plants—mosses and horsetails, violets and pansies—and young shoots that have just pushed up through the soil. Collect flowers, leaves, and seeds, but take only one of each kind. Add to your list of dinner guests the caterpillars, butterflies, and slugs you find on different plants. If a bird touches down just long enough to fly off with a twig or grass, it is probably busy building a nest.

Wind blows grass pollen from flower to flower.

Grass flower

Seeds fall from plants; are carried by the wind; and are picked up and dropped by ants, squirrels, and birds.

What looks and creeps like a simple animal, then grows a stalk and, like a mushroom, gives off spores? —Your small square yard slime mold.

Leaf Rubbing

Place a leaf between two sheets of paper. On a hard surface rub a crayon or a pencil back and forth over the upper sheet. Watch as a print of the leaf appears.

Leaves have many different shapes and sizes. Some are pointed at the tips, while others are rounded. Leaf edges can be smooth or notched. Make prints of all the kinds of leaves in your yard. Which are most alike? Which most different?

Growing Mold

Place a piece of bread or fruit in a jar. Then seal it with a lid. Leave it until you see a powdery gray or green fuzz on the food. The fuzz is mold. It grew from spores that were in the air and landed on the food.

After a few days, throw away the jar with the moldy food inside. Some molds produce helpful medicines. Others can make you sick.

Night Life

A chipmunk hurries back to its burrow. Birds disappear among branches. Squirrels and bees retreat to their nests. The last rays of the sun are fading as the chill of night overtakes the heat of the day. In the eerie twilight stillness, the yard suddenly seems deserted. Then flashes from a firefly tell a different story: The night shift is on the move. Those animals that spent the day resting or hiding are coming out into the open to take their turn at making a living.

Creatures of the night depend on sharpened senses to survive where there is little or no light. Most night animals, such as mice, can see much better than you can in very dim light. Extrasensitive whiskers help guide a mouse as it feels its way through the dark. High

To attract a mate, a firefly must blink the right pattern of colored light. It's fun to turn your collecting jar into a lantern by capturing fireflies. But release the animals soon, for they have work to do.

Most moths hide by day and fly by night. Try to spot a white-lined sphinx moth hovering as it sips sweet nectar.

Under cover of darkness, an earthworm pokes its head out of its burrow to find leaves. It keeps its tail in for a quick retreat.

Frightened by a skunk, young shrews hold on to each other and to their mother.

22

up in a tree, an owl hearing the faintest squeak can tell the mouse's exact location. From another direction, a skunk eating an earthworm picks up the scent of mouse and follows its nose toward more dinner. Just as the owl takes to the air, the mouse, startled by a moth, scurries to safety. It was unaware of the real danger—the owl and the skunk.

If you see a moth looping and diving like a plane at an air show, it isn't showing off but trying to save itself from a bat. From a distance the bat is sending out very high sounds to find dinner. If the sounds bounce off the moth and echo back to the bat, it can tell where the moth is. The bat then flies toward the insect like one missile intercepting another. The moth can hear the bat coming and tries to escape it. But more often than not, there is no escape.

Moths will be attracted to a flashlight, while other animals will be scared away. Try placing a red filter over the light. Since many animals can't see red, they may not know you are there.

With a five-inch wingspan, a black witch moth is as big as a small bird.

A tiger beetle searches in an owl pellet for remains of animals the owl ate.

Place an empty can in your square and see if daddy longlegs check in to spend the night.

Litter "Bugs"

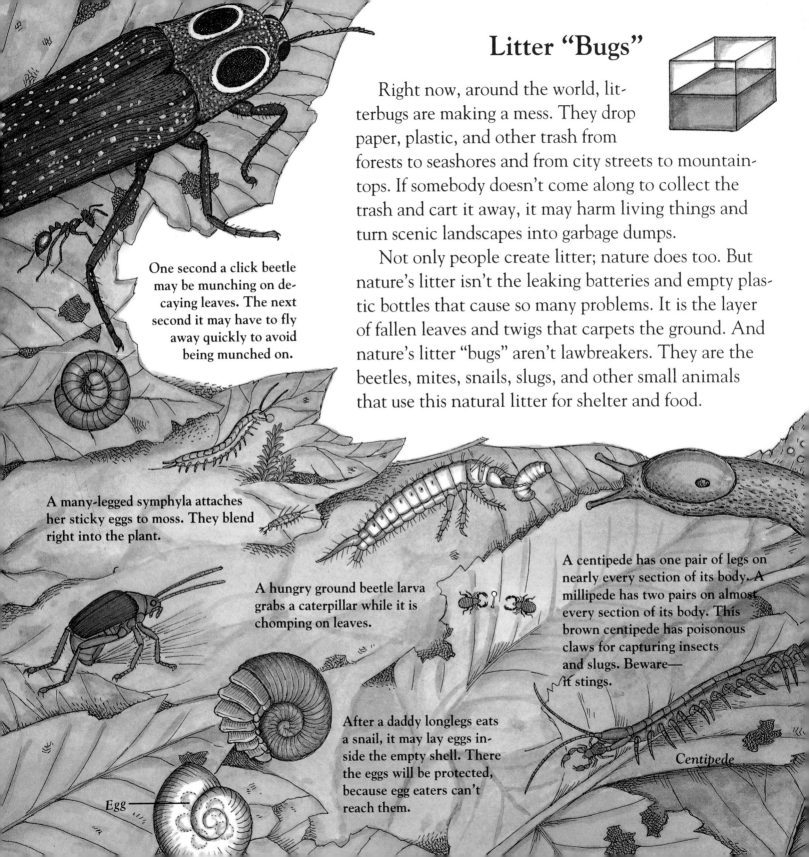

Right now, around the world, litterbugs are making a mess. They drop paper, plastic, and other trash from forests to seashores and from city streets to mountaintops. If somebody doesn't come along to collect the trash and cart it away, it may harm living things and turn scenic landscapes into garbage dumps.

Not only people create litter; nature does too. But nature's litter isn't the leaking batteries and empty plastic bottles that cause so many problems. It is the layer of fallen leaves and twigs that carpets the ground. And nature's litter "bugs" aren't lawbreakers. They are the beetles, mites, snails, slugs, and other small animals that use this natural litter for shelter and food.

One second a click beetle may be munching on decaying leaves. The next second it may have to fly away quickly to avoid being munched on.

A many-legged symphyla attaches her sticky eggs to moss. They blend right into the plant.

A hungry ground beetle larva grabs a caterpillar while it is chomping on leaves.

A centipede has one pair of legs on nearly every section of its body. A millipede has two pairs on almost every section of its body. This brown centipede has poisonous claws for capturing insects and slugs. Beware— it stings.

After a daddy longlegs eats a snail, it may lay eggs inside the empty shell. There the eggs will be protected, because egg eaters can't reach them.

Egg

Centipede

Lift some "litter" in your small square and feel how damp it is underneath. Litter dwellers find the cool, moist maze shaped by fallen leaves and twigs the ideal spot to make their living. There are hiding places in the litter protected from the drying heat of the sun. The leaves themselves are food for thousands of snails, slugs, millipedes, and insects that mostly feed at night. By breaking leaves apart, these plant eaters help funguses and other recyclers begin the work of getting rid of nature's litter. Nothing is wasted or harmed, and the soil is enriched.

There's no need for you to put leaf litter, animal droppings, or dead plants and animals you find in your square into a trash can. Nature's recycling team makes the best possible use of them.

Catch a Track

Among the best clues left behind by visitors to your yard are tracks. Tracks can help you quickly identify rabbits, mice, birds, and other animals that find their way to your square. You can catch tracks by covering a paper plate with soil, wetting the soil, and leaving the plate on the ground for animals to walk over. Make a sketch of each track left on your plate, and look for claw marks. For more about tracks, see page 29.

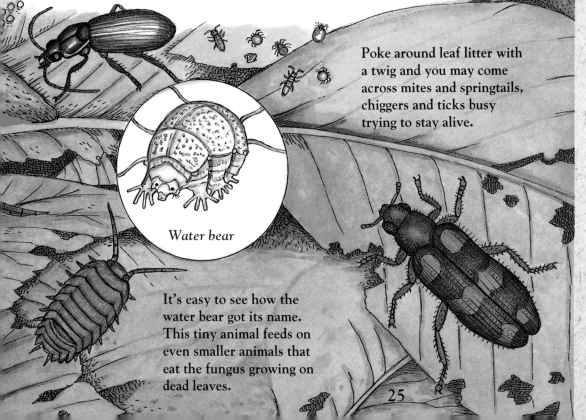

Poke around leaf litter with a twig and you may come across mites and springtails, chiggers and ticks busy trying to stay alive.

Water bear

It's easy to see how the water bear got its name. This tiny animal feeds on even smaller animals that eat the fungus growing on dead leaves.

Thousands of animals like those above feast on fallen leaves, droppings, and dead animals and plants. Use your magnifying glass to search for as many of these litter feeders as you can find. They are part of nature's recycling team, reusing what they can to grow and stay healthy. Be very careful with any American carrion beetle, for it is an endangered species.

Fungus threads feed on leaves and animal remains, taking in nutrients and freeing minerals. Litter "bugs" gobble up all they can eat, leaving their droppings as food for other animals. Insects lay eggs on droppings and animal remains so the young will find food waiting for them when they hatch. Birds carry off twigs, leaves, and mouthfuls of fur for building nests. In the meantime, recyclers too small to be seen except with a microscope move in to attack any and every part of the litter they can reach. In your small square there are millions of these bacteria and one-celled animals. They make their living by breaking down everything inside dead plants and animals into simple forms that living plants can use.

American carrion beetle

Bacteria

Protozoa

Funguses

Day after day the attack continues until the job is done. By then, the leaves, animals' bodies, and droppings have disappeared. Recyclers have produced dark, pudding-like humus rich in minerals and nutrients they freed from all the dead plants and animals. Now it is time for earthworms and other soil animals to take over by mixing the humus into the earth.

As the seasons change, look for decaying leaves in your square. Tie a string around a leaf stem and see how it changes each month. Because humus is always being mixed into your soil, you won't be able to tell where humus is. But each growing plant is a clue that the recyclers have been doing their work. Perhaps one day people will make such good use of all their litter that it won't mess up the world.

In the main attack force of nature's army of recyclers there are bacteria, one-celled animals, and funguses. Millions of each are at work decaying the litter in your square. Follow the arrows down to the animals that eat them.

Even in the litter, where there are plant eaters there are hunters. Predatory beetles, ants, and centipedes prowl for prey through the litter. Not only do they capture litter feeders, but they also eat other hunters such as springtails and beetle mites.

27

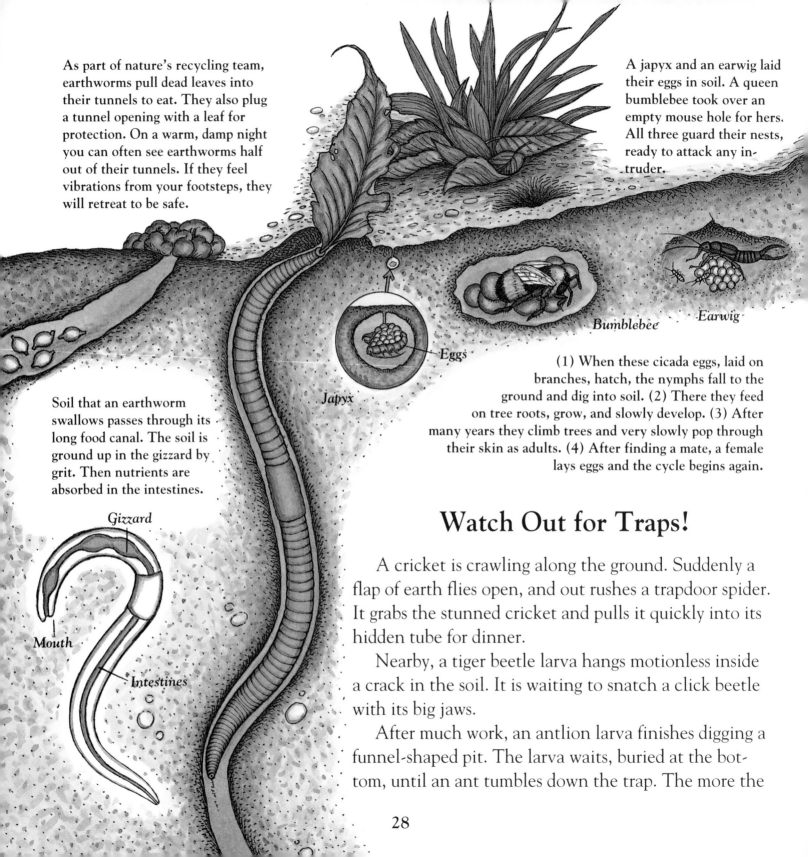

As part of nature's recycling team, earthworms pull dead leaves into their tunnels to eat. They also plug a tunnel opening with a leaf for protection. On a warm, damp night you can often see earthworms half out of their tunnels. If they feel vibrations from your footsteps, they will retreat to be safe.

A japyx and an earwig laid their eggs in soil. A queen bumblebee took over an empty mouse hole for hers. All three guard their nests, ready to attack any intruder.

Eggs

Bumblebee

Earwig

Japyx

Soil that an earthworm swallows passes through its long food canal. The soil is ground up in the gizzard by grit. Then nutrients are absorbed in the intestines.

Gizzard

Mouth

Intestines

(1) When these cicada eggs, laid on branches, hatch, the nymphs fall to the ground and dig into soil. (2) There they feed on tree roots, grow, and slowly develop. (3) After many years they climb trees and very slowly pop through their skin as adults. (4) After finding a mate, a female lays eggs and the cycle begins again.

Watch Out for Traps!

A cricket is crawling along the ground. Suddenly a flap of earth flies open, and out rushes a trapdoor spider. It grabs the stunned cricket and pulls it quickly into its hidden tube for dinner.

Nearby, a tiger beetle larva hangs motionless inside a crack in the soil. It is waiting to snatch a click beetle with its big jaws.

After much work, an antlion larva finishes digging a funnel-shaped pit. The larva waits, buried at the bottom, until an ant tumbles down the trap. The more the

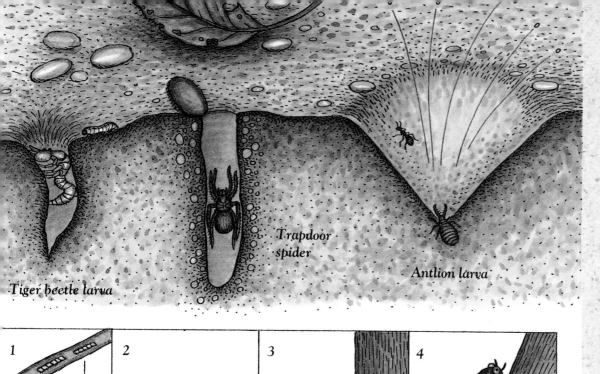

Tiger beetle larva

Trapdoor spider

Antlion larva

If you find tracks in mud, cut out a strip of cardboard about 3 or 4 inches high and as long as you need to fit around the tracks. Fasten the ends together with a paper clip to form a ring. Press the ring into the mud around the tracks, leaving about 2 inches sticking aboveground.

In a clean coffee can mix water and plaster of Paris until smooth and thick. Pour the plaster quickly into the ring and wait 15 to 20 minutes until hardened.

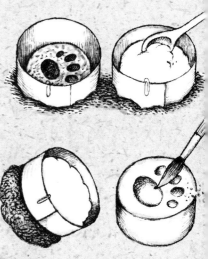

Carefully dig the plaster cast out, remove the ring, and wrap the cast in newspaper.

Leave it inside for a day. Then gently wash away any mud with water and a brush. Use the illustrations on page 25 or in a field guide to figure out what animal made the tracks.

ant tries to climb out, the more the dirt loosens under its legs. Finally the tired insect falls into the hungry mouth of the antlion.

Not all the holes in the ground of your square are traps like these. One may be the entrance to a mouse's nest. Another may be an exit for newly hatched Japanese beetles. And still another may contain a cicada that has been growing underground in your yard for many years. Most holes, though, are clues that your soil is being cared for by nature's soft, squirmy, slimy gardeners—earthworms.

29

Earthworms spend almost all their time underground. But they do slither through the litter to get leaves or dead insects to drag below. And they surface for air when heavy rains flood their tunnels. Underground they eat dirt, grinding it up to free nutrients from dead plant and animal bits. Earthworms change some soil chemicals into forms that plants can use to grow and stay healthy.

With every swallow, an earthworm eats out a skinny tunnel through which it can move. Some tunnels open at the surface, where an earthworm pushes out brown lumps of droppings called castings. Castings contain finely ground soil, food wastes, and the chemicals for plants.

Look for castings in your square. They are a sign that hundreds of earthworms are tunneling below, loosening your soil, breaking up clumps, mixing in humus, stirring in minerals, and making spaces for air and water that plants and soil animals must take in to stay alive.

Life underground for earthworms and other soil animals is just as difficult as life for surface dwellers. Instead of a tangle of leaves and stems to muddle through, there is a jungle of plant roots and fungus threads at every turn in the earth. Instead of fallen logs and spiderwebs, there are many rocks and buried seeds to avoid. And of course there are predators. From above, a hungry robin has its beak pointed toward the opening to an earthworm's tunnel. From below, the sharp teeth of a furry shrew are ready to capture yet another worm.

Eggs

Root cells

This grass plant has several main roots that branch and rebranch. They hold the plant in soil and absorb minerals and water. All its roots stretched end to end would measure hundreds of miles.

Sand

Space

Clay

Silt

Plant roots help hold soil together. This soil shown in the circle consists of sand, silt, and clay with spaces between.

Earthworms and moles loosen soil as they tunnel. They make passages for rain that soaks into the ground and for growing roots seeking water. Some roots reach more than 10 feet into the earth. Others stay near the surface, branching over a large area.

Iris stem

Tiger lily bulb

If a squirrel doesn't find all the acorns it buried, one or two of them may grow into new oak trees. Chipmunks and mice also hide seeds in the ground.

The iris stem that grows underground isn't safe from the hungry beetle feeding inside it. If you dig up a lily bulb in late summer, you may find a clue that some animal ate the rest of the plant.

Japyx

Dig and you might find a japyx capturing a pseudoscorpion.

Soil bacteria in the oval change nitrogen in air into a form that plants can use. Bacteria in the circle help make humus.

Sometimes funguses and plants are partners. The fungus soaks up nutrients from soil and passes them to plants' root cells. The roots pass along sugar in return.

Taproot

This nematode worm won't be eating any more soil bacteria, because it is trapped in the threads of a fungus.

Fungus

Dandelions have one main taproot, from which side roots branch. So do carrots and pine trees. Dig carefully to keep from harming roots.

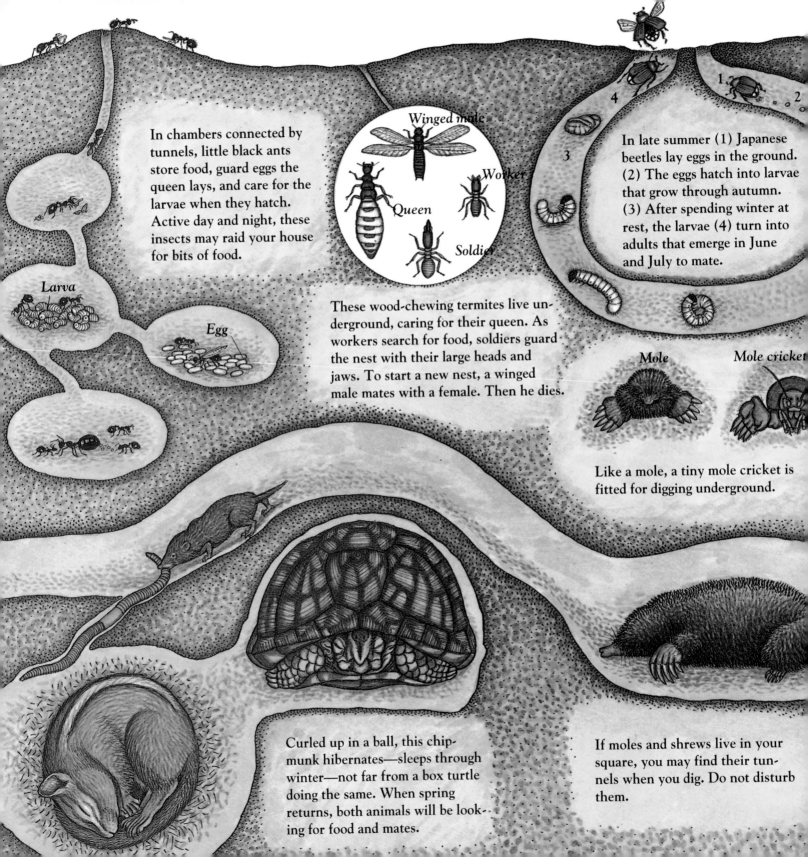

In chambers connected by tunnels, little black ants store food, guard eggs the queen lays, and care for the larvae when they hatch. Active day and night, these insects may raid your house for bits of food.

Winged mole

Queen

Worker

Soldier

In late summer (1) Japanese beetles lay eggs in the ground. (2) The eggs hatch into larvae that grow through autumn. (3) After spending winter at rest, the larvae (4) turn into adults that emerge in June and July to mate.

Larva

Egg

These wood-chewing termites live underground, caring for their queen. As workers search for food, soldiers guard the nest with their large heads and jaws. To start a new nest, a winged male mates with a female. Then he dies.

Mole

Mole cricket

Like a mole, a tiny mole cricket is fitted for digging underground.

Curled up in a ball, this chipmunk hibernates—sleeps through winter—not far from a box turtle doing the same. When spring returns, both animals will be looking for food and mates.

If moles and shrews live in your square, you may find their tunnels when you dig. Do not disturb them.

Torpedo-shaped moles are well equipped for digging. Their large clawed paws, powered by strong muscles, act like shovels. They loosen soil and scoop it to the side, pressing it into the walls of shallow, deep-branched tunnels. Moles push leftover soil up to the surface and leave small piles called molehills. If you find one in your square, you can be sure a mole is eating, digging, resting, and perhaps nesting below.

Like other creatures that live where there is little or no light, most moles are blind. They find their way through the earth with the touch of sensitive snout bumps, whiskers, and body hairs. If a mole finds more earthworms than it can eat, it bites off their heads so they cannot escape. Then it stores them in a tunnel chamber so they are fresh for a later meal. There is so much food for moles underground that they don't surface often. It's worth noting if you see one aboveground.

Who's There?

Dig out a hole in the soil a couple of inches deep, and place the dirt in your collecting jar. Cover your tray with a screen, and turn the jar over onto it. Position your flashlight as shown. Then turn it on.

To escape the heat and light, soil animals will fall into the tray. Examine them with your glass, draw pictures in your book, then return the creatures to the earth. Pour the soil back into the hole.

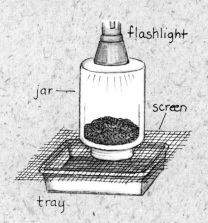

flashlight
jar
screen
tray

Cyst

When the weather gets warmer and wetter, the copepod breaks out of its hard protective cyst.

A one-celled animal called an amoeba keeps changing its shape as it moves through the film of water that surrounds soil particles.

Amoeba

A Worm in the Hand

Pick up an earthworm and feel the slimy mucus that covers it and the tiny bristles that help anchor it in its tunnel. Make a sketch, noting how the body is divided into ringlike sections. Place the worm back where you found it

33

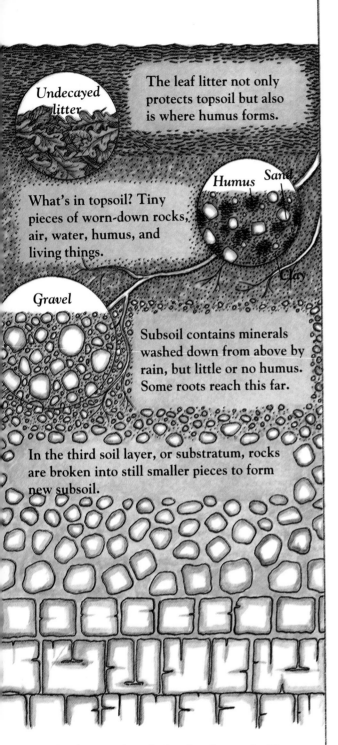

Undecayed litter

The leaf litter not only protects topsoil but also is where humus forms.

Humus   Sand

What's in topsoil? Tiny pieces of worn-down rocks, air, water, humus, and living things.

Clay

Gravel

Subsoil contains minerals washed down from above by rain, but little or no humus. Some roots reach this far.

In the third soil layer, or substratum, rocks are broken into still smaller pieces to form new subsoil.

The deeper you dig, the harder the earth. Underneath your soil there is solid rock (which you probably won't reach) that is part of Earth's crust. The three soil layers above this rock are also called (from the surface down) the A, B, and C horizons.

More Than Just Dirt

The next time you sit down to dinner, take a good look at everything in front of you. All the plant parts on your plate were nourished by soil. The meat came from animals that ate plants that grew in soil. If your table is made of wood, it came from trees that grew in soil. Without soil there would be no food, no table to put it on, and no living things in your small square.

Dig down in your small square and scoop out a handful of soil. It took hundreds, perhaps thousands, of years for the rocks to be broken down into the bits and pieces of sand, silt, and clay that you are holding in your hand. The different ways nature does this are called weathering.

Try to find tree roots growing out of cracks in rocks. The roots are breaking the rocks open by widening the cracks as they grow. This is weathering. Fill an ice tray with water and let it freeze. In the same way that water expands in the tray when it forms ice, water filling cracks in rocks expands on very cold nights and pushes cracks open farther. Over time, with repeated freezing and thawing, the cracks keep getting bigger until the rocks break. This is weathering, too.

Keep digging. Within 18 inches you may find three soil layers. The dark top layer is topsoil. The lighter-colored middle layer is subsoil. The third layer, where rocks are still being broken down to form new soil, is the substratum. In these three layers, the nonliving world of rocks and minerals helps the living world stay alive.

Need water? Turn on a faucet. Need food? Go shopping. Flick a switch, and electricity flows. Turn a dial, and oil burns for heat. No trucks or pipes make deliveries to your small square, yet nature provides the necessities of life without running out.

Energy comes from the sun, but just about everything else is recycled. Take rain that soaks into the ground and is absorbed by roots. It rises to leaves to be used for making food and keeping cells working. What plants can't use is released into the air as the gas called water vapor. As air cools, water vapor changes into droplets that form clouds and more rain. The way water is used and reused is called the water cycle.

Follow the arrows of the food chains to find out which animals eat which plants or other animals. When living things die, their bodies decay, and plants reuse the nutrients.

Oak

Sparrow hawk

Garter snake

Toad

Grasshopper

Roots absorbing nutrients and water

Decayers

Dig It

Choose a spot, lay down some newspaper, and start digging. Examine each shovelful of soil on the newspaper. Note in your book what it is, the color, if it is rough or smooth, and so on. Is it alive with creatures? Which? Does it contain pebbles or leaves?

Measure where the soil color changes. Then draw a picture of your soil similar to the one on the opposite page. Be sure to replace all soil and animals so the animals are safe from predators.

Soil Shake

In your collection jar, mix about 2 inches of topsoil with water. Cover the jar, shake, and allow to stand. Watch as the largest soil particles, gravel, settle on the bottom, followed by sand, then silt, then clay. Floating on top are humus and litter not fully decayed.

water
humus
clay/silt
sand
gravel

Bring the Outside In

Find a clear plastic box, a jar, or a fish tank. Cover the bottom with about 1/2 inch of gravel. Then layer about 1/2 inch of charcoal on top. (You can buy charcoal at a store that sells tanks.) Now cover the charcoal with 1 to 3 inches of topsoil you dig from under a rock or log in your yard.

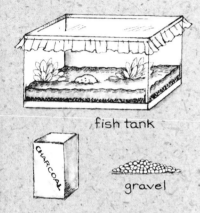

fish tank

CHARCOAL

gravel

Carefully dig up some small plants, and replant them in the soil. Near the plants, place rocks or pebbles for soil animals to hide under. Cover the container with a piece of glass, clear plastic, or plastic wrap. Place it where there is light but not direct sun. If the soil is dry, water it. The cover will keep in moisture, and the plants will make food and oxygen as they use up waste carbon dioxide.

Nature in Action

Something new is always happening in your small square. Day and night, season to season, year after year, change keeps taking place above and below ground.

The more time you spend getting to know your square, the more you will witness nature in action. By figuring out how each living thing is fitted to survive where it does; how plants, animals, soil, and air are connected; and how nature provides for life to go on, you will deepen your understanding of why so many different kinds of creatures exist on this, the only planet in the universe known to have food makers, oxygen breathers, egg layers, and dirt eaters.

There are more than a million reasons for you to take care of your square, and every one of them lives in it. Keep your square free of bug killers (pesticides) and other harmful chemicals. Don't hurt anything on purpose. In these ways you will be letting nature select which plants and animals live, which die, and which produce young. These are the same kinds of steps people can take in parks, on farms, in rain forests, on mountainsides—everywhere on Earth—to help keep our planet healthy.

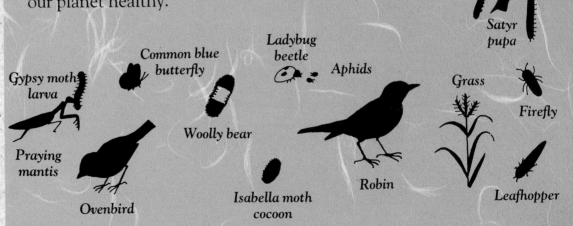

Clover

Blister beetle

Deer track

Skunk track

Maple seed

Elm seed

Tiger beetle

Honeybee

Millipede

Satyr pupa

Grass

Firefly

Leafhopper

Gypsy moth larva

Common blue butterfly

Ladybug beetle

Aphids

Woolly bear

Praying mantis

Ovenbird

Isabella moth cocoon

Robin

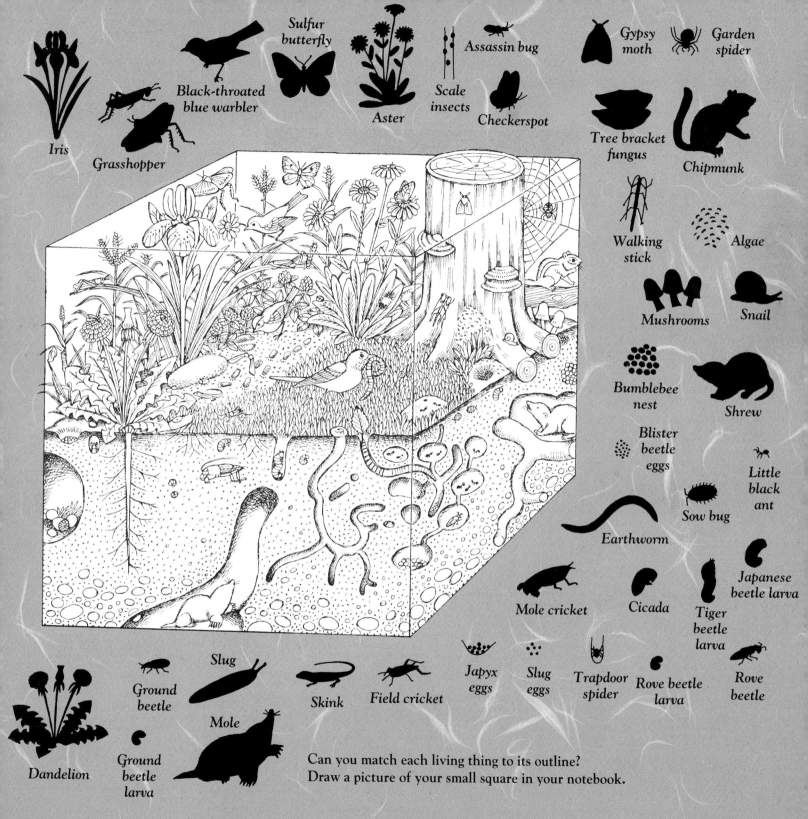

Iris

Grasshopper

Black-throated blue warbler

Sulfur butterfly

Aster

Assassin bug

Scale insects

Checkerspot

Gypsy moth

Garden spider

Tree bracket fungus

Chipmunk

Walking stick

Algae

Mushrooms

Snail

Bumblebee nest

Shrew

Blister beetle eggs

Little black ant

Sow bug

Earthworm

Mole cricket

Cicada

Tiger beetle larva

Japanese beetle larva

Rove beetle

Rove beetle larva

Trapdoor spider

Slug eggs

Japyx eggs

Field cricket

Skink

Slug

Ground beetle

Mole

Ground beetle larva

Dandelion

Can you match each living thing to its outline?
Draw a picture of your small square in your notebook.

Which of these vertebrates (animals with back-bones) appear in your small square all year long, and which just visit during certain seasons?

Only mammals grow fur. Only birds grow feathers. Most reptiles have dry, scaly skin. Amphibians have thin, scaleless, moist skin.

Mammals

Raccoon

White-footed mouse

Norway brown rat

Least shrew

Eastern mole

Little brown bat

Gray squirrel

Prairie vole

Cottontail rabbit

Eastern chipmunk

Fence lizard

Reptiles and Amphibians

Box turtle

Whitetail deer

Striped skunk

Birds

Sparrow hawk

Screech owl

Bobwhite

Chickadee

Downy woodpecker

Brown thrasher

Ruby-throated hummingbird

Robin

Hermit thrush

Blackpoll warbler

Ovenbird

Black-throated blue warbler

Garter snake

Five-lined skink

Spring peeper

American toad

39

None of these backyard animals has bones in its body. They are all invertebrates. Most of the invertebrates in your square are insects or other arthropods.

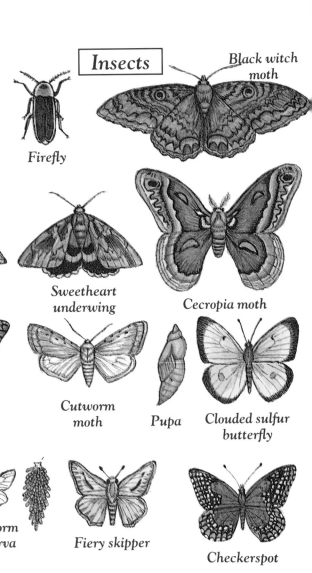

Insects

Firefly

Black witch moth

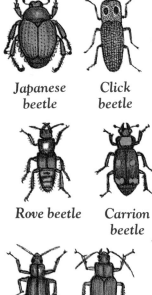
Japanese beetle

Click beetle

Blister beetle

Checkered beetle

Locust underwing

Sweetheart underwing

Cecropia moth

Rove beetle

Carrion beetle

Meadow fritillary

Zerene fritillary

Cutworm moth

Pupa

Clouded sulfur butterfly

Ground beetles

Tiger beetle

Hermes satyr

Bagworm and larva

Fiery skipper

Checkerspot

Bombardier beetle

Ladybug beetle

Borer moth

Tussock moth

Isabella moth and larva

Praying mantis

Engraver beetle

Antlion

Scale insect

Leafhopper

Termite

Mole cricket

Walking stick

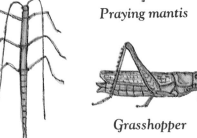 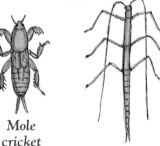
Grasshopper

40

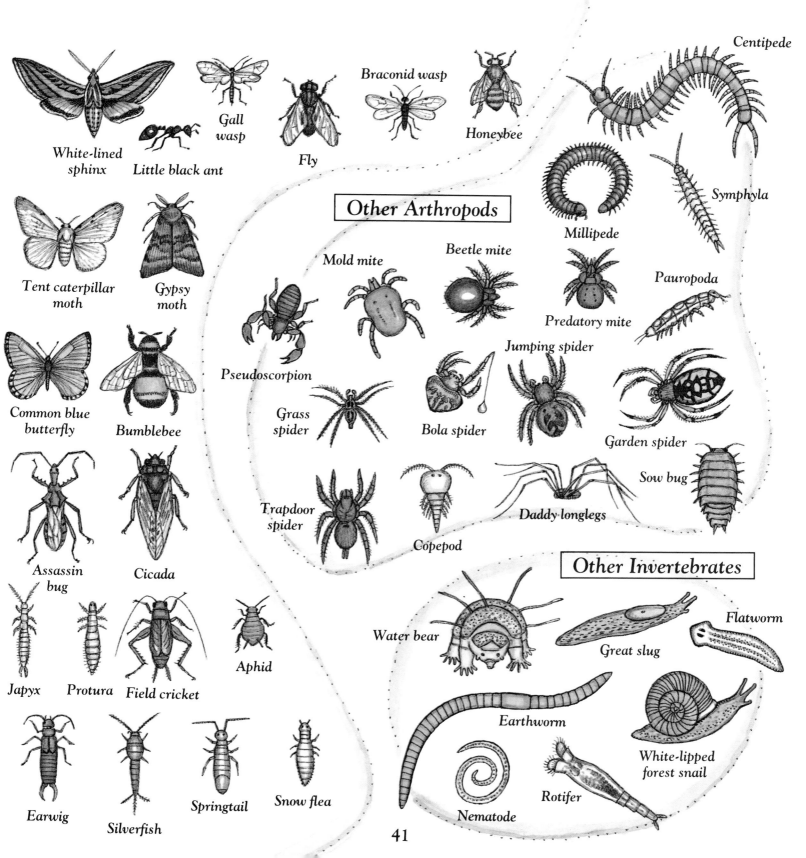

White-lined sphinx

Little black ant

Gall wasp

Fly

Braconid wasp

Honeybee

Centipede

Tent caterpillar moth

Gypsy moth

Common blue butterfly

Bumblebee

Assassin bug

Cicada

Japyx

Protura

Field cricket

Aphid

Earwig

Silverfish

Springtail

Snow flea

Other Arthropods

Mold mite

Beetle mite

Millipede

Symphyla

Pauropoda

Pseudoscorpion

Predatory mite

Jumping spider

Grass spider

Bola spider

Garden spider

Trapdoor spider

Copepod

Daddy-longlegs

Sow bug

Other Invertebrates

Water bear

Great slug

Flatworm

Earthworm

White-lipped forest snail

Nematode

Rotifer

41

Are any of these plants or funguses growing in your square? To see one-celled soil creatures or tiny funguses, you will need a microscope. Don't forget to look at rocks. They too are an important part of nature.

Plants

White oak

Grass

Aster

American elm

Maple

White clover

Black-eyed Susan

White pine

Acacia

Shagbark hickory

Dandelion

Horsetail

Moss

Violet

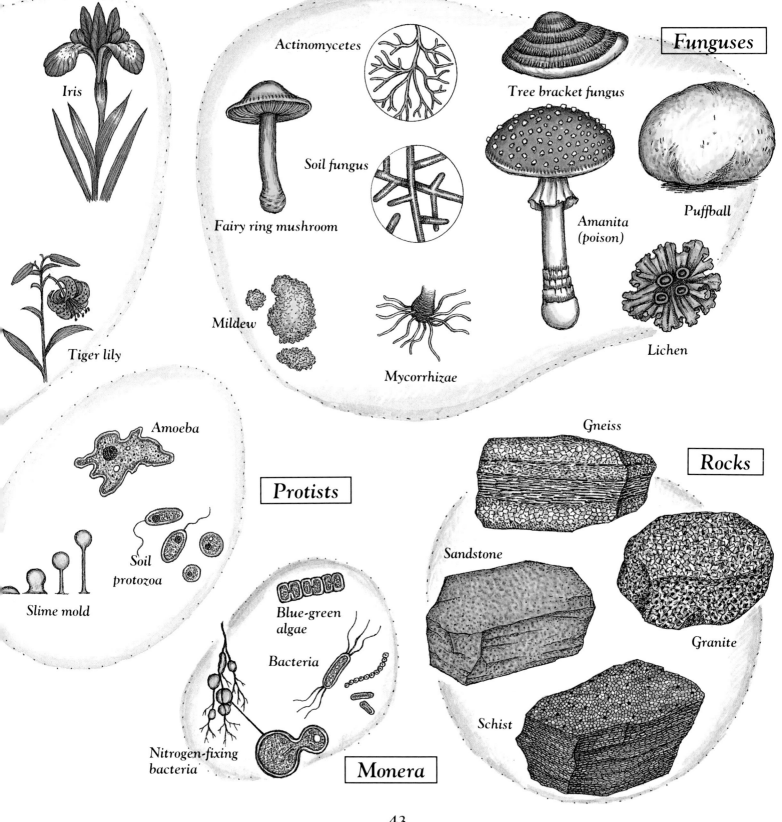

Iris

Tiger lily

Actinomycetes

Fairy ring mushroom

Soil fungus

Mildew

Mycorrhizae

Tree bracket fungus

Amanita
(poison)

Puffball

Lichen

Funguses

Amoeba

Soil
protozoa

Slime mold

Protists

Blue-green
algae

Bacteria

Nitrogen-fixing
bacteria

Monera

Gneiss

Sandstone

Granite

Schist

Rocks

43

Index

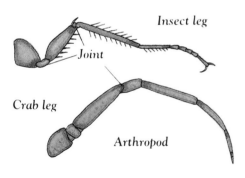
Chrysalis

Mourning cloak butterfly
Alfalfa butterfly
Checkerspot butterfly

Cocoon

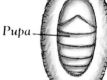

Pupa
Outside
Inside

Index

Invertebrate

Amoeba

Snail

Honeybee

Earthworm

Egg

Insect

Reptile

Amphibian

Bird

Metamorphosis

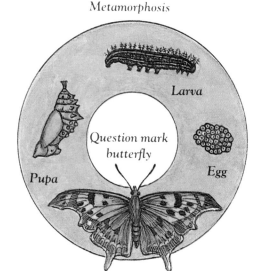

Larva

Egg

Pupa

Question mark butterfly

Index

Owl pellet

Teeth

Bones

Fur

Skull

Feather

Sparrows

White-crowned   Chipping   Song

White-throated

Index

Vertebrate

Backbone

Bird

Amphibian

Mammal

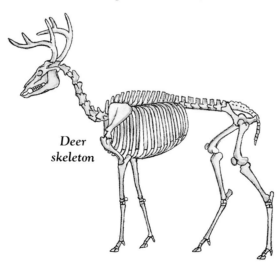

Deer skeleton

Further Reading

To find out more, look for the following in a library or bookstore:

Golden Guides, Golden Press, New York, NY

Golden Field Guides, Golden Press, New York, NY

The Audubon Society Beginner Guides, Random House, New York, NY

The Audubon Society Field Guides, Alfred A. Knopf, New York, NY

The Peterson Field Guides, Houghton Mifflin Co., Boston, MA

Reader's Digest North American Wildlife, Reader's Digest, Pleasantville, NY

Look in an art supply store or in the library for books on how to draw plants and animals. If you like to sketch and paint outdoors, here are some things you'll find handy:

paper

number 2 pencil

paintbrush

bottle of black ink

tray of watercolors

eraser

plastic bottle for water

stiff cardboard or clipboard to draw on